Optimizing Fleet Performance with Data Analytics

Table of Contents

Preface

Introduction

Prelude

Chapter 1: Understanding the Basics of Data Analytics

Chapter 2: Data Collection and Integration

Chapter 3: Key Metrics for Fleet Performance Optimization

Chapter 4: Leveraging Predictive Analytics for Maintenance

Chapter 5: Route Optimization with Data Analytics

Chapter 6: Enhancing Fleet Safety with Data Analytics

Chapter 7: Future Trends in Fleet Performance and Data Analytics

Conclusion

Glossary

Preface

The maritime and offshore industries are undergoing rapid transformation, driven by advancements in technology, stricter regulatory demands, and the increasing need for enhanced safety, operational efficiency, and compliance. Recognizing the industry's evolving needs, the Gosships Learning Series was developed to provide maritime professionals with practical, accessible, and targeted knowledge to stay competitive in this changing landscape.

This series is designed to offer foundational to intermediate knowledge with a focus on real-world applications. Each book in the series includes a certification test, ensuring that the information gained is not only understood but can be effectively applied to daily operations.

The Gosships Learning Series empowers maritime and energy industry professionals, from entry-level crew members to shoreside managers, equipping them with the tools necessary to navigate the complexities of modern fleet operations. We hope that this series will contribute to your professional development and open new doors for growth and success in your career.

Introduction

Welcome to the Gosships Learning Series, designed for professionals eager to expand their knowledge and advance their careers in maritime and energy sectors. This book, titled *Optimizing Fleet Performance with Data Analytics*, has been meticulously developed by industry experts to provide comprehensive insights into the role of data analytics in improving fleet efficiency and performance. Whether you're new to the subject or looking to deepen your understanding, this book is tailored to help you stand out as an expert in your field.

In this book, we will explore the following key areas:

- **Data Collection Techniques**: Understand how data is gathered from vessels through sensors, monitoring systems, and IoT devices.

- **Performance Analysis**: Learn how to interpret and analyze operational data to identify inefficiencies and optimize vessel performance.

- **Predictive Maintenance**: Discover the benefits of predictive analytics in forecasting equipment failures, minimizing downtime, and reducing costs.

- **Dynamic Route Optimization**: Explore how real-time data can help optimize shipping routes, reduce fuel consumption, and improve delivery times.

- **Environmental Impact**: Delve into the role of data analytics in monitoring emissions and helping companies meet environmental compliance goals.

After reading this book, you will be prepared to take an assessment that evaluates your comprehension of the material. Upon successfully passing the assessment, you will receive a Certificate of Achievement, available at www.gosships.com, validating your expertise in optimizing fleet performance with data analytics.

Who Is This Book For?

This book is designed for:

- **Maritime professionals** seeking to improve their understanding of data-driven fleet management.

- **Fleet managers** looking to enhance their ability to optimize operations using data analytics.

- **Aspiring students** wanting to enter the maritime or logistics industries with a solid understanding of modern technologies.

- **Government and regulatory personnel** aiming to stay informed about the latest advancements in data-driven fleet performance.

By mastering the concepts outlined in this book, you'll be equipped to tackle the challenges of modern fleet management, ensure compliance with international standards, and contribute to more efficient, safer maritime operations.

Thank you for choosing the Gosships Learning Series to support your ongoing journey of professional growth and continuous learning. We are confident that the knowledge gained through this resource will help you advance in your career and improve the performance of the fleets you manage.

Gosships Learning Series 2024/2025

1. Hydrogen: The Fuel of the Future
2. Green Ammonia: The Next Big Thing in Shipping
3. Decarbonizing Shipping: Pathways to Zero Emissions
4. Battery Technology for Industrial Applications
5. Carbon Capture and Storage: Can It Save the Planet?
6. Biofuels 101: Turning Waste into Energy
7. Understanding LNG (Liquefied Natural Gas)
8. Methanol as a Marine Fuel
9. Offshore Wind Energy: The Future of Renewable Power
10. Tidal and Wave Energy: Harnessing the Ocean
11. Electrofuels: The Next Generation of Carbon-Neutral Fuels
12. Energy Storage Systems for Grid Reliability
13. Hydrogen Fuel Cells for Transportation
14. Solar Energy Innovations: Beyond Solar Panels
15. Smart Grids: The Backbone of Future Energy Systems
16. Ammonia-Hydrogen Blends: A Dual Fuel Solution?
17. Nuclear Power: Small Modular Reactors for a Low-Carbon Future
18. Hydropower: The Oldest Renewable Energy Source
19. Decentralized Energy Systems: Microgrids for Resilience
20. Energy Efficiency Technologies for Industry
21. Hydrogen Production from Seawater
22. Fuel Cells for Maritime Applications
23. Geothermal Energy: Unlocking Earth's Heat
24. Future of EV Charging Infrastructure
25. Synthetic Fuels: Bridging the Gap to Decarbonization
26. Cybersecurity for Maritime and Offshore Operations
27. AI and Automation in Shipping and Logistics
28. Digital Twins in Maritime: Revolutionizing Asset Management

29	Risk Management in Offshore and Maritime Operations
30	Compliance with IMO 2020 Regulations
31	Sustainable Ship Design: Reducing Environmental Impact
32	Marine Renewable Energy: Wave, Tidal, and Offshore Wind Integration
33	Ballast Water Management Systems
34	Blockchain Technology in Shipping: Improving Transpc'y & Efficiency
35	Effective Supply Chain Management for Energy Industries
36	Leadership in the Energy Transition
37	Effective Crisis Management in Maritime Operations
38	Shipyard Safety Management Systems
39	Port State Control (PSC) Inspection Readiness
40	Remote Vessel Operations and Autonomous Shipping
41	Optimizing Fleet Performance with Data Analytics
42	Maritime Environmental Regulations: Staying Ahead of Compliance
43	Advanced Maintenance Strategies: Condition Monitoring & Predictive Maintenance
44	Global LNG Market: Trends and Opportunities
45	Incident Investigation in Maritime Operations
46	International Maritime Law: Key Concepts and Applications
47	Emergency Preparedness and Response for Offshore Oil & Gas
48	Energy Transition Strategies for Oil and Gas Companies
49	Maritime Drones: Applications and Safety Considerations
50	Effective Project Management in Offshore Energy Projects

All Rights Reserved Disclaimer

The contents of this book, including but not limited to all text, graphics, images, logos, and designs, are the intellectual property of Gosships LLC and are protected by copyright law. No part of this publication may be reproduced, distributed, transmitted, displayed, or modified in any form or by any means, including photocopying, recording, or other electronic or mechanical methods, without the prior written permission of the publisher, except in the case of brief quotations in critical reviews or articles.

The information contained within this book is for educational purposes only and is provided "as is" without warranty of any kind, either expressed or implied. The authors and publishers disclaim any liability for any direct, indirect, or consequential loss or damage arising from the use of the material in this book.

For permissions or inquiries, please contact: admin@gosships.com

© 2024 Gosships LLC. All rights reserved.

Prelude

Fleet performance optimization is no longer just a competitive edge in today's maritime industry—it is a necessity. With growing demands for efficiency, sustainability, and cost-effectiveness, maritime companies are increasingly turning to data analytics to gain insights that drive smarter decision-making. The application of data analytics in fleet management helps operators monitor and improve every aspect of vessel performance, from fuel consumption to maintenance schedules, safety practices, and route optimization.

Data analytics transforms raw data from vessels into actionable insights, offering a way to monitor real-time performance, predict equipment failures, and reduce operational costs. By harnessing the power of data, fleet managers can optimize vessel utilization, reduce downtime, and make proactive decisions that lead to improved safety and sustainability.

This mini-book is designed for beginners and those with intermediate knowledge of fleet management and data analytics. We will explore the fundamentals of data analytics, its key applications in fleet management, and the future of data-driven fleet operations. This guide will also include real-world examples, case studies, and supporting charts to help visualize the benefits and practical applications of data analytics in the maritime industry.

Chapter 1

Understanding the Basics of Data Analytics

What is Data Analytics?

Data analytics refers to the process of examining large sets of data to uncover patterns, correlations, and trends that can help inform decision-making. In the context of fleet management, data analytics involves collecting data from various sources—such as vessel sensors, historical records, and external data like weather conditions—and analyzing it to optimize operations.

Fleet managers can use data analytics to understand vessel performance, identify inefficiencies, and anticipate maintenance needs. There are different types of analytics used in fleet performance optimization:

- **Descriptive Analytics**: This involves analyzing past performance to understand trends and performance metrics. It answers the question, "What happened?" by analyzing historical data.

- **Predictive Analytics**: This type of analytics uses data, statistical algorithms, and machine learning techniques to predict future outcomes. It answers the question, "What is likely to happen?" by identifying patterns that can indicate future performance or failures.

- **Prescriptive Analytics**: This takes predictive analytics a step further by providing recommendations on actions to take based on data insights. It answers the question, "What should we do?" by offering actionable strategies to optimize operations.

Chart: Types of Data Analytics in Fleet Management

Type of Analytics	Key Focus	Example Use Case
Descriptive Analytics	Understanding past performance	Analyzing fuel consumption trends over the last year
Predictive Analytics	Forecasting future performance	Predicting when an engine will need maintenance
Prescriptive Analytics	Recommending actions based on data	Suggesting route adjustments to reduce fuel usage

Data Sources for Fleet Management

The success of data analytics in fleet performance depends on the quality and variety of data collected. Common data sources include:

- **Telematics and IoT Sensors**: Vessels are equipped with Internet of Things (IoT) sensors that monitor engine performance, fuel consumption, and other key operational metrics in real-time.

- **External Data**: Weather reports, port congestion information, and traffic patterns can be integrated with onboard data to help optimize route planning and operational efficiency.

- **Historical Data**: Data from past voyages, maintenance records, and operational performance can be analyzed to forecast trends and anticipate future needs.

Chapter 2
Data Collection and Integration

Key Data Collection Points in Fleet Operations

Data collection is critical to understanding and improving fleet performance. Onboard telematics systems collect vast amounts of data from various parts of the vessel, including:

- **Engine and Machinery Performance**: Sensors monitor critical components like engines, pumps, and generators. This data helps detect inefficiencies or mechanical issues before they escalate.

- **Fuel Usage and Emissions Tracking**: Monitoring fuel consumption is essential for cost management and environmental compliance. Data analytics can track how much fuel is being used, where inefficiencies occur, and how emissions can be minimized.

- **Crew Performance and Onboard Safety**: Data on crew behavior and adherence to safety protocols is essential for ensuring that vessels operate safely and within regulatory frameworks.

Data Integration Across Fleet Management Systems

One of the major challenges fleet operators face is the integration of data from multiple systems. Many fleets operate with different software platforms, making it difficult to consolidate data into a single view for analysis. However, with the advent of cloud-based platforms and advanced data management systems, fleets can now integrate data from multiple sources into a single, centralized system for analysis.

- **Real-Time Data Integration**: Using IoT sensors, fleet managers can gather real-time data from vessels. This data can be accessed and analyzed immediately, enabling quick decision-making.

- **Historical and Real-Time Data Synergy**: Combining real-time data with historical data can provide powerful insights. For example, historical data on fuel consumption can be compared to real-time performance to detect inefficiencies.

Graphic: Integrated Data Flow in Fleet Management

- **Sensors onboard vessels** collect data ☐ Data transmitted to **cloud-based platforms** ☐ **Data analysts and fleet managers** access and analyze data ☐ **Real-time decisions** improve operations.

Tools for Data Collection

The tools used to collect and process fleet data are becoming more sophisticated as technology evolves. Some key tools include:

- **Telematics Systems**: These systems use GPS and onboard sensors to provide real-time data on location, speed, fuel consumption, and more.

- **IoT Platforms**: IoT sensors placed throughout the vessel collect data on engine performance, temperature, pressure, and other key metrics. These sensors transmit data to cloud-based platforms for analysis.

Chapter 3

Key Metrics for Fleet Performance Optimization

Fuel Efficiency Metrics

Fuel costs represent one of the largest expenses in maritime operations, so optimizing fuel efficiency is crucial. Data analytics allows fleet operators to track and optimize the following metrics:

- **Fuel Consumption per Voyage**: This metric tracks the total fuel used during a trip, helping fleet managers compare performance across different routes and vessels.

- **Fuel Efficiency Based on Load and Weather Conditions**: Vessels consume fuel differently based on their load, speed, and the weather. Data analytics can help identify optimal operating conditions to minimize fuel consumption.

- **Impact of Speed Optimization**: Ships that travel at high speeds often consume more fuel. By analyzing speed data in relation to fuel usage, fleet managers can determine the most fuel-efficient speeds for different routes.

Chart: Fuel Consumption vs. Speed Optimization

Speed (knots)	Fuel Consumption (liters/hour)	Efficiency (Fuel/km)
10	200	High
15	350	Moderate
20	600	Low

Maintenance and Downtime Metrics

Predictive maintenance allows fleet managers to reduce unplanned downtime by analyzing data on engine wear and other key performance indicators. The following metrics are crucial:

- **Predictive Maintenance Indicators**: Metrics such as oil quality, engine temperature, and vibration levels can help predict when a component is likely to fail.

- **Maintenance Costs per Vessel**: Tracking the cost of repairs and maintenance helps identify the most expensive aspects of fleet operations, which can then be optimized.

- **Downtime**: By analyzing when and why vessels are out of operation, fleet managers can identify ways to minimize downtime and improve vessel utilization.

Operational Efficiency Metrics

Operational efficiency metrics provide insights into how well the fleet is being utilized:

- **Vessel Utilization Rates**: This measures the percentage of time a vessel is in use versus idle. High utilization rates typically indicate better performance.

- **Route Optimization**: Data analytics can compare different routes to determine which ones are the most efficient in terms of fuel consumption, time at sea, and delivery schedules.

- **On-Time Performance**: Tracking how often deliveries arrive on time can help improve customer satisfaction and operational efficiency.

Chapter 4
Leveraging Predictive Analytics for Maintenance

What is Predictive Maintenance?
Predictive maintenance is an advanced maintenance strategy that utilizes data analytics and real-time monitoring to predict equipment failure before it happens. This proactive approach allows for repairs or replacements to be carried out before a malfunction occurs, reducing the risk of unplanned downtime and minimizing costly equipment damage. Predictive maintenance is more efficient than both reactive maintenance—where issues are only addressed after a breakdown—and preventive maintenance, which follows a fixed schedule that may not accurately reflect the actual condition of the equipment.

Graphic: Predictive Maintenance Process

1. **Data Collection**: Sensors are installed on equipment to monitor various operational parameters in real-time, such as temperature, pressure, and vibration.

2. **Data Analysis**: Advanced algorithms analyze the data to identify patterns, deviations, or anomalies that may indicate an impending failure.

3. **Maintenance Alert**: When the system detects signs of potential failure, it triggers an alert, allowing the crew to take preventive action before the equipment breaks down.

4. **Maintenance Performed**: Maintenance is scheduled and performed based on data-driven insights, ensuring timely repairs and avoiding unexpected downtime.

How Predictive Analytics Works in Fleet Management
In fleet management, predictive analytics enhances the efficiency and reliability of vessel operations by forecasting equipment failures based on both historical data and real-time sensor data. This approach can be applied to critical components such as engines, turbines, and generators, which are vital to the smooth functioning of a ship.

- **Historical Data**: Past maintenance records, performance data, and the operational history of each piece of equipment are analyzed to determine the typical lifespan and failure patterns of components. This data helps create a baseline for predicting when future repairs might be needed.

- **Real-Time Data**: Sensors continuously monitor the performance of key machinery, collecting data on crucial parameters such as temperature, pressure, vibration, and fuel consumption. These real-time insights allow fleet operators to detect signs of wear or malfunction early, enabling them to schedule maintenance tasks at the optimal time.

Benefits of Predictive Maintenance

Predictive maintenance in maritime operations offers several benefits, including:

- **Reduced Downtime**: By addressing potential failures before they lead to breakdowns, predictive maintenance minimizes unplanned downtime, keeping vessels in operation and reducing delays.

- **Lower Costs**: Preventing equipment failure reduces the need for costly emergency repairs and replacements. It also helps optimize the use of spare parts and minimizes labor costs by streamlining maintenance efforts.

- **Increased Safety**: Regular monitoring of equipment health through predictive analytics enhances onboard safety by ensuring that critical systems are functioning optimally at all times.

- **Extended Equipment Lifespan**: By preventing wear and tear from escalating into more significant issues, predictive maintenance prolongs the useful life of machinery and equipment, contributing to overall fleet efficiency.

Predictive maintenance is transforming fleet management by providing actionable insights, enabling smarter decision-making, and enhancing operational reliability.

Chapter 5
Route Optimization with Data Analytics

How Data Analytics Improves Route Planning

Route optimization is one of the key benefits of applying data analytics in fleet management. By analyzing various factors such as weather patterns, fuel consumption, and traffic congestion, data analytics can help fleets identify the most efficient routes for each voyage.

- **Real-Time Data Integration**: By incorporating real-time weather data and port conditions, fleets can avoid delays and reduce fuel consumption.
- **Predictive Analytics for Route Planning**: Using predictive analytics, fleet managers can foresee potential disruptions such as storms or port congestion and adjust routes in advance.

Benefits of Dynamic Route Optimization

- **Fuel Savings**: Optimizing routes can significantly reduce fuel consumption by selecting the shortest and most efficient paths.
- **Reduced Delays**: Dynamic routing allows ships to avoid congested ports and bad weather, reducing the risk of delays and improving on-time performance.

Case Study: Data-Driven Route Optimization

A global shipping company implemented a data-driven route optimization system that used real-time analytics to adjust routes based on weather conditions and port congestion. As a result, the company reduced fuel consumption by 12% and improved on-time performance by 20%. This case study highlights the impact of data analytics on improving fleet efficiency and reducing operational costs.

Chapter 6
Enhancing Fleet Safety with Data Analytics

Safety Analytics in Fleet Management
In the maritime industry, safety is paramount, and the use of data analytics has revolutionized how fleet managers ensure the safety of vessels, crew, and cargo. By collecting and analyzing data from various sources such as equipment performance, crew behavior, and environmental conditions, fleet managers can identify risks and potential hazards before they escalate into accidents. This proactive approach to safety management allows for better decision-making and quicker responses to emerging threats.

1. **Crew Behavior Monitoring**
   One of the critical aspects of fleet safety is ensuring that crew members adhere to established safety protocols. Data analytics can track crew behavior in real-time, monitoring whether they are wearing the necessary safety equipment, following proper procedures, and adhering to regulations during high-risk operations. For example, wearable devices can provide data on crew fatigue, which is a significant factor in maritime accidents. By analyzing crew performance and activity levels, fleet managers can intervene before fatigue becomes a safety issue, scheduling breaks or making adjustments to shifts accordingly.

2. **Predictive Safety Analytics**
   Predictive analytics, an advanced form of data analysis, allows fleet managers to anticipate potential risks. By analyzing patterns in equipment performance or navigation errors, predictive models can forecast when accidents might occur. For instance, if certain engine components have a history of failing after a specific number of operating hours, predictive analytics can alert the crew and management when those components are likely to fail, allowing for preemptive maintenance. Similarly, by analyzing navigation patterns, predictive analytics can detect potential collision risks or navigation errors that could lead to accidents, giving crews enough time to adjust their course.

Using Data to Improve Vessel Safety

Data analytics offers several ways to enhance the safety of maritime operations. Onboard sensors continuously collect data from critical systems such as engines, navigation systems, and environmental sensors. This data is analyzed in real-time, enabling fleet managers and crews to respond quickly to potential safety threats.

1. **Real-Time Alerts**
   Sensors onboard vessels constantly monitor essential systems such as engine performance, fuel levels, hull integrity, and environmental conditions. If the sensors detect any anomalies, such as an overheating engine, excessive fuel consumption, or a vessel straying off course, real-time alerts are sent to the crew. These alerts allow the crew to take immediate corrective actions, such as adjusting the throttle or rerouting the vessel to avoid hazardous weather. In this way, real-time data helps mitigate risks before they escalate into serious incidents.

2. **Predictive Models**
   In addition to real-time monitoring, predictive models can forecast potential safety risks by using historical data to identify patterns that might indicate future failures. For example, historical data on equipment performance, such as engine wear or fuel efficiency trends, can reveal when a failure is likely to occur. By using predictive models, fleet managers can schedule maintenance or take preventive measures to avoid breakdowns or other accidents, ensuring the safe and continuous operation of the vessel.

Chapter 7

Future Trends in Fleet Performance and Data Analytics

AI and Machine Learning in Fleet Analytics

Artificial intelligence (AI) and machine learning (ML) are transforming fleet performance optimization. AI can analyze vast amounts of data from sensors, logs, and other sources, providing real-time insights that help fleet managers make more informed decisions. Machine learning, in particular, allows AI systems to learn from past experiences, improving the accuracy and effectiveness of predictions over time.

1. **AI-Driven Predictive Analytics**
   With AI-driven predictive analytics, data from onboard sensors is processed at incredible speeds, identifying trends and patterns that human operators may miss. For example, AI can track the wear and tear of key mechanical components and predict the ideal time for maintenance, preventing unscheduled breakdowns and reducing costs. This kind of real-time insight is invaluable for improving fleet safety, as it allows for quicker responses to safety issues and more efficient management of resources.

Big Data and Maritime Operations

The maritime industry generates massive amounts of data, often referred to as "big data." This data includes everything from engine performance metrics to weather forecasts and shipping traffic patterns. Advanced analytics platforms can process and analyze these large datasets to provide actionable insights, helping fleet operators optimize vessel performance and safety.

1. **Cloud-Based Big Data Platforms**
   Cloud-based platforms aggregate data from multiple sources and vessels, making it easier for fleet managers to access and analyze data in real-time, regardless of geographic location. This enables global fleet management, allowing operators to monitor multiple vessels simultaneously, predict maintenance needs, and optimize routes based on the latest data. Cloud-based platforms also enable collaboration between different stakeholders in the maritime industry, ensuring that everyone has access to the same information for decision-making.

The Rise of Digital Twins in Fleet Operations

A digital twin is a virtual model of a physical vessel that simulates real-time operations. These digital replicas allow fleet operators to monitor vessel performance, predict maintenance needs, and simulate how different operational scenarios would affect the ship.

1. **Simulation of Fleet Operations**
   Digital twins provide operators with a powerful tool for simulating various fleet operations, such as route optimization, fuel efficiency, and safety management. For instance, operators can test different navigation routes to determine the safest and most fuel-efficient path for a vessel, reducing fuel consumption and improving operational efficiency. Additionally, digital twins allow fleet managers to simulate maintenance scenarios, predicting the impact of equipment failure and testing potential solutions without putting the actual vessel at risk.

Conclusion

The use of data analytics in maritime fleet management is becoming a cornerstone of modern operations. By leveraging data from onboard sensors, historical records, and external sources, fleet managers can enhance safety, improve efficiency, and reduce operational costs. Predictive models, AI, and machine learning are key technologies driving these advancements, allowing fleet operators to anticipate problems before they occur and optimize fleet performance.

As the maritime industry evolves, data analytics will play an increasingly critical role in enhancing vessel safety, optimizing performance, and minimizing environmental impact. The adoption of advanced analytics techniques such as predictive maintenance, dynamic route optimization, and AI-driven decision-making is essential for staying competitive in an industry that is constantly under pressure to improve efficiency and safety while reducing costs.

Moreover, the integration of big data and digital twin technologies promises to revolutionize how fleets are managed, offering unparalleled insights into vessel performance and enabling smarter, more informed decision-making. By investing in these technologies, fleet managers can not only improve operational performance but also ensure the long-term sustainability and safety of their fleets.

Glossary: Optimizing Fleet Performance with Data Analytics

1. **AI (Artificial Intelligence)**: Technology that simulates human intelligence, used in data analysis and decision-making to optimize fleet performance.

2. **API (Application Programming Interface)**: A toolset that allows software systems to communicate and share data, enabling integration of fleet management systems.

3. **Big Data**: Large datasets collected from sources like sensors and IoT devices, analyzed to improve fleet operations and identify trends.

4. **Blockchain**: A decentralized digital ledger used in shipping to securely track transactions and cargo, ensuring transparency and reducing fraud.

5. **Cloud Computing**: The delivery of computing services over the internet, allowing real-time fleet data storage and analysis from anywhere.

6. **Compliance Reporting**: Automated generation of reports to ensure fleet operations adhere to international regulatory standards.

7. **Crew Behavior Monitoring**: Data analysis of crew activities and adherence to safety protocols, contributing to safer and more efficient fleet management.

8. **Cybersecurity**: The protection of data collected from fleet management systems against unauthorized access and cyber threats.

9. **Dashboard Analytics**: Visual representation of key performance data on a dashboard, providing fleet managers with real-time insights.

10. **Data Analytics**: The examination of raw data to draw conclusions and optimize fleet performance in areas like maintenance and fuel efficiency.

11. **Data Collection**: The process of gathering information from sensors and systems onboard vessels to monitor and improve performance.

12. **Data Governance**: The management and control of data quality, security, and compliance within fleet analytics to ensure accuracy and ethical usage.

13. **Data Integration**: Combining data from various sources to provide a unified view of fleet operations, aiding in decision-making.

14. **Data Mining**: The process of analyzing large datasets to uncover

patterns and insights that can improve fleet performance.

15. **Digital Twin**: A virtual model of a vessel, used to simulate and analyze real-time performance and predict maintenance needs.

16. **Dynamic Route Optimization**: Real-time analysis of shipping routes using data to choose the most efficient paths, reducing fuel consumption and costs.

17. **Efficiency Metrics**: Measurements used to assess vessel performance, such as fuel efficiency, speed, and maintenance intervals.

18. **Environmental Compliance**: Ensuring fleet operations meet environmental regulations, such as emissions standards, monitored via data analytics.

19. **Fleet Efficiency**: A measure of how effectively a fleet operates, optimizing resources such as fuel and maintenance to improve overall performance.

20. **Fleet Management System (FMS)**: Software used to monitor and manage fleet operations, including routing, maintenance, and performance tracking.

21. **Fleet Optimization**: Continuous improvement of fleet operations through data-driven decisions to maximize performance and minimize costs.

22. **Fuel Consumption**: The amount of fuel used by a vessel, analyzed through data to reduce usage and improve operational efficiency.

23. **Geofencing**: Using GPS data to create virtual boundaries for vessel operations, ensuring ships stay within designated areas.

24. **Historical Data**: Previously collected fleet performance data used to identify trends and make predictions for future operations.

25. **Internet of Things (IoT)**: A network of connected devices, such as sensors, that collect and transmit data to monitor vessel performance.

26. **Key Performance Indicators (KPIs)**: Metrics that evaluate the effectiveness of fleet performance, such as maintenance efficiency and fuel usage.

27. **Latency**: The delay between data collection and analysis, which can affect real-time decision-making in fleet management.

28. **Machine Learning (ML)**: A subset of AI that allows systems to learn from data patterns and improve predictive analytics for fleet performance.

29. **Maintenance Alert**: A notification generated by fleet systems to inform operators of potential equipment failures, allowing preemptive action.

30. **Maintenance Optimization**: Using predictive analytics to forecast equipment failures and schedule repairs before downtime occurs.

31. **Operational Efficiency**: The effectiveness of fleet operations, measured by metrics such as fuel consumption, maintenance frequency, and delivery speed.

32. **Performance Dashboard**: A digital interface that displays real-time data on fleet performance, helping managers monitor and improve operations.

33. **Predictive Analytics**: The use of data models to forecast future events, such as equipment failures, helping to make proactive decisions.

34. **Predictive Maintenance**: A maintenance strategy that uses data analytics to predict when equipment will fail, allowing repairs to be made before breakdowns occur.

35. **Real-Time Alerts**: Immediate notifications generated by fleet systems in response to safety risks or performance anomalies.

36. **Real-Time Data**: Information collected and processed instantly, providing real-time insights into fleet performance and safety.

37. **Remote Monitoring**: Tracking fleet operations and vessel performance from a remote location using real-time data and communication technologies.

38. **Route Optimization**: The process of analyzing data to determine the most efficient shipping routes, reducing costs and improving delivery times.

39. **Safety Analytics**: The use of data to identify potential safety risks and ensure compliance with regulations in fleet operations.

40. **Sensor Data**: Information gathered from sensors on vessels that monitor conditions like engine temperature, speed, and fuel levels.

41. **Simulation**: Using digital twins and other models to test fleet operations and make decisions without real-world risks.

42. **Sustainability**: The ability of a fleet to minimize its environmental impact, often measured through fuel efficiency and emissions tracking.

43. **Telematics**: The use of telecommunications technology to transmit data from vessels to a central system, enabling real-time monitoring.

44. **Uptime**: The duration a vessel remains operational without experiencing downtime, an important metric for measuring fleet performance.

45. **Vessel Monitoring System (VMS)**: A system used to track a vessel's location, speed, and performance in real-time, aiding in fleet management.

46. **Wearable Devices**: Technology worn by crew members to monitor their physical condition, helping ensure safety during maritime operations.

47. **Weather Data Integration**: Incorporating real-time weather forecasts into fleet management systems to adjust routes and ensure safety.

48. **Workload Balancing**: Distributing operations across a fleet to maximize efficiency and prevent overuse of individual vessels.

49. **Yield Optimization**: Adjusting fleet operations to maximize cargo delivery efficiency while minimizing fuel and resource consumption.

50. **Zero-Emission Technologies**: Innovations aimed at eliminating harmful emissions from vessels, often integrated with data analytics for sustainability.

www.ingramcontent.com/pod-product-compliance
Lightning Source LLC
Chambersburg PA
CBHW030041230526
45472CB00002B/626